SCOTLAND'S PAST IN ACTION

Building
Railways

James L Wood

N·M·S

NATIONAL MUSEUMS OF SCOTLAND

Published by the National Museums of Scotland
Chambers Street, Edinburgh EH1 1JF

ISBN 0 948636 82 3

© Trustees of the National Museums of Scotland 1996

British Library Cataloguing in Publication Data

A catalogue record for this book is available from the
British Library

Series editor Iseabail Macleod

Designed and produced by the Publications Office of
the National Museums of Scotland

Printed on Huntsman Velvet 110gsm by
Clifford Press Ltd, Coventry, Great Britain

Acknowledgements

We are grateful to the National Museums of Scotland Charitable Trust for
support for this publication.

Illustrations: Front cover, 57, 59, 61: Science and Society Picture Library.
Back cover, 4, 9, 12, 13, 14, 19, 22, 28, i,ii top and bottom, iv top and bottom,
v, vi top and bottom, viii, 42, 53, 55, 63, 70, 78: National Museums of
Scotland. 6: Charles Meacher. 8: Strathkelvin District Libraries and
Museums. 10: Glasgow Museums and Art Galleries. 29, 35, 41, 45: The
Mitchell Library, Glasgow. 39, 50, 74: Eric Simpson. iii top: J L Wood. iii
bottom: John Burnett. vii: Angus and Patricia Macdonald. 69: Hulton
Deutsch Collection. 73: Christine Wilson. 75: John Hume.

Front cover: *LNER poster of the Forth Bridge.* H G Gaw

Back cover: *Leaderfoot viaduct, near Melrose.* Ian Larner

CONTENTS

INTRODUCTION

This is a book about the creation of Scotland's railways and their operation, with the emphasis on the people involved. It is concerned mainly with nineteenth- and early twentieth-century developments but also deals briefly with how, in more recent times, they have been adapted to meet changing transport needs. In the physical sense, the creation of the railway system was the work of navvies, masons and others who actually built them but, in addition, many people were concerned with promotion, finance, design and supervision of the construction. Once built they had to be maintained and operated. Furthermore, because of the very nature of a railway system and its impact on the country, both economic and environmental, governments of all political colours became involved in the regulation of construction and operation, and this required a staff of civil servants.

When looking at colourful steam locomotives and carriages in museums or as models or even simply in old photographs it is very easy to believe that the railways existed mainly to allow these beautiful old trains to be displayed! Of course this was never the case even though railways in general, and steam railways in particular, have a very special place in the affections of many people. Railways were built because it was believed that they would meet a need for the transport of goods and people, and that in so doing they would be profitable. Sometimes they were profitable, occasionally very profitable, but many were bitter disappointments to their hopeful promoters and shareholders.

While the main concern is with Scottish railways, these cannot be totally separated from the British system. Scottish companies

Ballochmyle viaduct, near Mauchline, on the former Glasgow & South Western main line south of Kilmarnock. The central arch has a span of 181ft (55.2m). Completed in 1848, the viaduct is still in daily use. D O Hill

5

Thornton shed, 1920. The arrival of the brand new locomotive,
Glen Gloy, *is clearly an event. In addition to the driver and
fireman, the party includes the shedmaster, running foreman
and his son, and an inspector.* SEA

had substantial track mileage in the north of England, although
the converse was not the case. There was significant involvement
of English promoters and financiers and although Scotland has a
separate legal system, generally speaking, legislation relating to
the railways applied throughout the United Kingdom.

Although the railway system is not as extensive as it once was
there is still a lot to see. Much of the nineteenth-century infrastructure remains in everyday use, a testament to the skills of
those who built it, and there are also important surviving features
on routes which have been disused for many years. Many items of
historical interest can be seen in museums and original documents can be studied in record offices and libraries. The story of
the creation of Scotland's railways is a fascinating one. For
readers seeking more detailed information there is a brief bibliog-
raphy and some suggestions for things worth looking at.

BUILDING RAILWAYS

1 Communications before the Railways

Eighteenth-century Scotland saw the start of the major developments in trade and industry which have come to be known as the Industrial Revolution. The beginnings of this process were perceptible even before the Union of the Parliaments in 1707, and became increasingly apparent as the century passed. Agriculture became more productive, and factory-based manufacturing industries using power-driven machinery began, slowly at first, to replace the old 'cottage industries' and hand processes. The population grew and there was a movement from rural areas to the expanding towns where most of the new factories were located.

At the beginning of the eighteenth century roads were few and far between, and those which existed were poor. Until adequate roads had been constructed movement of both people and goods was difficult in summer and often impossible in winter. There were old tracks used by cattle drovers; pack-horses were used for goods but wheeled vehicles of any sort were rare. Travellers rode or walked according to their means. A few had their own coaches but while these advertised the status and wealth of the owners they were neither fast nor comfortable. The easiest way to move both goods and people was by water. In 1757 the only public conveyance by land between Glasgow and London was a stage waggon which took three weeks for the journey.

Much was done in the second half of the eighteenth century to upgrade existing roads and build new ones. In parallel with the road improvements public transport services developed. By 1783 there was a coach service between Edinburgh and London fifteen times a week, taking four days for the journey. In 1784 the first of the mail coaches started running between London and Bristol and soon there was an extensive network of services throughout the country. The coaches were provided under contract to the Post

Office. They were primarily for the carriage of mail, under the supervision of an armed guard who was a Post Office employee, but passengers were also carried. In 1788 a mail coach service between Glasgow and London, via Moffat, was started, taking 66 hours. Road improvements enabled this to be reduced until by 1836 the journey time was 42 hours from London to Glasgow and 46 hours in the reverse direction. Horses were changed every ten miles or so, and the process took no more than about a minute. Passengers had only three refreshment stops on the road, each of 35 to 40 minutes. A winter journey under these conditions must have been horrendous and quite unlike the glamorous image conveyed in numerous Christmas card scenes of coaches outside quaint inns or dashing through snow-covered countryside.

All the work carried out to improve the road system did little to aid the transport of heavy goods, such as coal. For these, water transport by river and coastal shipping remained by far the cheapest and most convenient method, where it was possible. With the development of industry, especially mining, there was a need to extend the benefits of cheap transport by water to areas away from the coast. The answer was the construction of artificial waterways. By comparison with England, Scotland had few canals, but they were nevertheless important aids to industrial development.

This 1830s view if the Monkland & Kirkintilloch Railway at Gartsherrie ironworks shows a horse-drawn train but from 1831 steam locomotives were sometimes used.

The most important canals were the linked trio in central Scotland, the Forth & Clyde, Monkland and Edinburgh & Glasgow Union. Connecting Grangemouth on the Forth with Bowling on the Clyde, the Forth & Clyde Canal was opened piecemeal between 1773 and 1790. The Monkland, from near Airdrie to Glasgow, was also completed in stages between 1773 and 1793. The third canal, the Edinburgh & Glasgow Union, linking Edinburgh with the Forth & Clyde at Falkirk, was completed in 1822. Many new collieries, ironworks, chemical works and other enterprises were established along their banks. Although primarily carriers of the raw materials and products of industry, there was also a significant volume of passenger traffic. In the Highlands the importance of water transport was fully appreciated. As a result, the Caledonian Canal was built through the Great Glen at government expense. It was completed in 1822. Making use of Loch Linnhe, Loch Lochy and Loch Ness this created a water route from Fort William to Inverness, thereby providing a safer passage for ships than the alternative route round the north of Scotland. The Crinan Canal across the Kintyre peninsula was promoted by local landowners. It opened in 1801 but subsequent major repairs required government financial assistance and in 1848 it was taken over completely. There

Mining activity around Clackmannan, showing the waggonway which took coal to the Forth for shipment. From an early nineteenth-century watercolour.

were various other proposals to build canals in the late eighteenth and early nineteenth century but few of these came to anything.

In view of the importance of transport by water in the eighteenth century it is not surprising that the first railways built in Scotland were designed to facilitate the carriage of coal from the pits to the nearest navigable water. Primitive railways were in use as adjuncts to the sixteenth-century German mining industry and they had spread to England by the early seventeenth century. The first of these 'waggonways' in Scotland was opened in 1722 to take coal from the collieries near Tranent to the harbours at Cockenzie and Port Seton, on the River Forth. Apart from being the first railway in Scotland this waggonway has a further claim to fame, in that the battle of Prestonpans in 1745 was fought across part of its route. By the end of the century most of the major ports on the Forth were connected to local pits to facilitate the shipment of coal. Notable examples were the extensive network of lines leading to Alloa, which came into use from about 1768 onwards, the system leading to Limekilns (1768) and the Fordell waggonway (about 1752) leading to St Davids harbour. These

St Davids harbour, terminus of the Fordell waggonway, shortly before closure of the line in 1946.

waggonways were originally equipped with wooden rails and worked by horse power. In time, iron replaced wood for the rails, and locomotives were used instead of horses. The longest lived was the Fordell waggonway, which did not close until 1946.

2 Development of the network

The Scottish railway network was created by about 200 companies which were, in the beginning, independent. Sometimes this independence was more apparent than real, with the smaller companies having been set up with the encouragement and assistance of their larger neighbours and the expectation of eventual union. For the most part, individual railways were planned simply on the basis of an expectation of profit or to thwart the schemes of rival companies. In Britain, unlike many European countries, there was little or no planning of the system as a whole. Although created by 200 enterprises the railways of Scotland were, by a complex sequence of amalgamations, take-overs and working arrangements, under the effective control of five companies from about 1870 onwards. The Caledonian Railway was dominant in Glasgow and the west of Scotland but had interests in the Edinburgh area. The Glasgow & South Western was active in the area indicated by its title. The North British was centred in Edinburgh, with interests in Glasgow and the west, and virtual monopolies in Fife and the south east. Between the three companies there were large overlaps and fierce competition for some of the traffic. For example, all three provided rail and ship services to the Clyde coast resorts and competed strongly for the custom of holidaymakers and commuters. The Highland Railway operated north and west of Perth, extending as far as Wick and Thurso, and Kyle of Lochalsh. The Great North of Scotland, the only 'Great' railway in Scotland, was in fact the smallest of the five. It operated throughout the north east and had a degree of overlap at its western side with the Highland, which caused a certain amount of friction.

The early nineteenth-century railways were more elaborate versions of the colliery waggonways of the previous century. Some passenger services were operated and steam locomotives came into use but their main purpose was still the same, the transport of coal to a navigable waterway. The Kilmarnock & Troon, opened in 1812 and the Monkland & Kirkintilloch, which began operations in 1826, are examples. Both were running passenger services and experimenting with steam haulage within a few years of opening.

The decade of the 1830s witnessed the final stages of the transition from the private waggonway into the full-blown public railway, conceived as a transport system in its own right rather than as an adjunct to a waterway. At the same time steam locomotives steadily replaced horse power. Virtually all the stages in this process of change were exhibited in the lines opened in the single year of 1831. Still dependent on its link to a waterway was the Ardrossan & Johnstone Railway which started at the end of the Glasgow, Paisley & Johnstone Canal. The Edinburgh & Dalkeith, still horse-worked, had no waterway connection. It was built to carry coal to Edinburgh although there was also a passenger service. The Dundee & Newtyle had three steep inclines in the course of its route. At the start horses were used on the level sections between the inclines while stationary steam winding engines hauled the trains up the inclines themselves. Locomotives came into use on the level sections in 1833. By far

Stuck in a snow drift near Thurso. Snow could cause major operating problems for railways, especially in the Highlands. SEA

the most significant of the 1831 railways was the Garnkirk & Glasgow. Its prime purpose was to bring coal, and other minerals, into Glasgow. Instead of being a feeder to a waterway it was in direct competition with the Monkland canal. Steam locomotives, from Robert Stephenson & Company of Newcastle, were used for both goods and passenger trains. The official opening was on 27 September 1831 although some trains, both horse and steam hauled, were run some months earlier.

The railways opened in the 1830s were largely planned in the 1820s. In 1830 the first inter-city trunk railway in Britain was opened between Liverpool and Manchester and its success greatly influenced the planning of the next generation of railways. Those planned in the 1830s were altogether grander and more ambitious than the fruits of 1820s thinking. Three major railways, opened in 1840-42, mark the real start of the age of the railway in Scotland. The engineering works, viaducts, embankments and cuttings needed to provide the easy gradients and gentle curves necessary for high-speed running, were on a much greater scale than anything previously seen on a Scottish railway. They were,

Artist's impression, about 1843, of one of the stations on the Dundee & Newtyle Railway. This is possibly the original terminus at Newtyle, showing the engine-house and chimney of the stationary steam winding engine used to haul trains up the incline in the background. SEA

Excavating a cutting. There are navvies are working at several different levels. The excavated material will be loaded into the waggons on the temporary track at the bottom of the excavation and taken away to form an embankment elsewhere.

of course, correspondingly more costly to build. The Glasgow, Paisley, Kilmarnock & Ayr Railway was opened in 1840 and the Glasgow, Paisley & Greenock in the following year. The third was the Edinburgh & Glasgow Railway, opened in 1842.

By the end of the decade the Anglo-Scottish trunk lines were in place. From 1848 it was possible to travel by train from London to Glasgow and Edinburgh, via Carlisle, Beattock and Carstairs. From Carlisle, the service was provided by the Caledonian Railway and associated companies. By 1850 Aberdeen and Dundee could also be reached by train from the south. For passengers to Glasgow and Edinburgh there was now even a choice of route. Travellers from London to Edinburgh could use the east-coast line by Newcastle, travelling from Berwick to Edinburgh by North British Railway. West-coast passengers for Glasgow could go via Dumfries and Kilmarnock, on what was to become the Glasgow & South Western Railway, as an alternative to the line by Beattock. Yet another Anglo-Scottish route became available in 1862 when the long North British branch line from Edinburgh to Hawick, opened in 1849, was extended to Carlisle.

While dealing with 'cross-border' routes the Irish connections should also be mentioned. The route from Dumfries to Stranraer was completed in 1861 and extended to Portpatrick in the following year. Although the improvement of communications with Ireland had been a strong incentive to build the line, establishing a satisfactory sea link proved difficult. There were three attempts in the 1860s but none lasted more than a few months. It was not until 1872 that a permanent service between Stranraer and Larne was started. The Glasgow to Ayr line of 1840 was extended to Girvan in 1860 and linked to Portpatrick and Stranraer in 1877.

Even within Scotland there was sometimes a choice of routes as early as 1850, for example from Edinburgh to Perth and Dundee. There was an all-rail route by the Glasgow line as far as Polmont and thence by Falkirk (Grahamston), Larbert and Stirling. The competing route, by the Edinburgh, Perth & Dundee Railway, involved a train journey to Granton, ferry to Burntisland, train to Perth via Ladybank. Dundee passengers travelled on by train to Tayport from where there was a steamer to Broughty Ferry, and finally yet another train to Dundee.

With the railway at Aberdeen by 1850, the next obvious destination to attract the attention of budding railway promoters was clearly the capital of the Highlands, Inverness. Promoted largely from Aberdeen, the Great North of Scotland Railway was authorized in 1846 but there was great difficulty in financing the project. After ten years the railway had been built only as far as Keith. Because of the slow progress a line was promoted from the Inverness end. The Inverness & Nairn Railway was opened in 1855 and extended to Keith in 1858, as the Inverness & Aberdeen Junction Railway. This made a through journey by rail from Aberdeen to Inverness possible for the first time. Possible it may have been but it was not easy at first, as the two companies involved did not make any great effort to cooperate with one another. The people of Inverness were not happy and wanted a direct line to the south. This was completed in 1863, as the Inverness & Perth Junction Railway from Forres, on the Inverness to Aberdeen line, via Aviemore and Drumochter summit.

Although the basic skeleton of the Scottish railway network was in place by the middle of the nineteenth century, a great number of lines remained to be built to flesh out the skeleton. The most significant increase in route mileage occurred in the cities, with many costly new lines. The cities themselves were expanding and rail traffic, both passenger and goods, was increasing rapidly. In Glasgow the main station for south-bound and Clyde-coast trains was in Bridge Street, south of the River Clyde. This handled both Caledonian and Glasgow & South Western Railway traffic. The location was inconvenient for many of the passengers and the station was too small for the growing traffic, with little room for expansion. The first railway to cross the Clyde within Glasgow was a joint North British and Glasgow & South Western line, promoted as the City of Glasgow Union Railway. This was designed to link the two companies' lines and make through services possible from north east to south west, across the city. Local services started in 1871 and in the following year through trains from Edinburgh to Greenock, and to Ayr,

1845. The only railways shown are the main lines in central Scotland and isolated lines round Dundee. There is no connection with England.

1852. There is now a network linking the major cities and most sizeable towns, with three routes into England.

1952. Although some closures took place from the 1930s onwards, the major reductions took place after 1952. This therefore shows the railway system virtually at its peak.

1985. This is in effect the present day network. Apart from the lines in the Highlands, there is a striking similarity between the pattern of routes still in use and those open in 1852.

Based on maps in *The Railway Surveyors* by Gordon Biddle

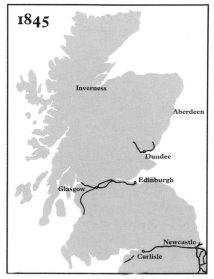

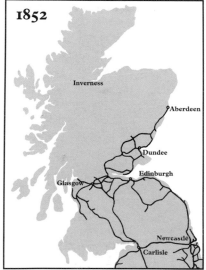

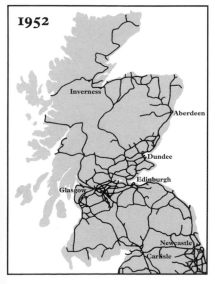

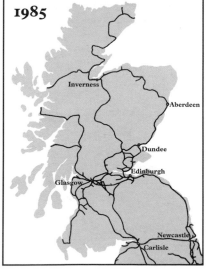

were introduced. As part of the project, the Glasgow & South Western's new terminus at St Enoch was completed in 1876. In the same year, the Midland Railway opened the Settle & Carlisle line. In cooperation with the Glasgow & South Western and North British this provided another route from London to Glasgow and, via the Waverley line, to Edinburgh. In 1879 the Caledonian opened Central Station, north of the Clyde, but such was the increase in traffic that a major extension soon became necessary. This was completed in 1905.

Major works in the city centre were inevitably difficult and expensive because of the cost of acquiring land and demolishing buildings. The City of Glasgow Union was welcomed by the local authority because it required the demolition of a large amount of slum property, but this was a special case. The North British and the Caledonian had similar problems on the north side of the river in that they had various lines which had been built piecemeal and had inconvenient connections between them. In addition the dock system was developing, resulting in increased rail traffic. Suburban passenger traffic was on the increase too, as new middle-class suburbs were built away from the city-centre grime. New lines running through the city centre were needed to tie the various pieces of the jigsaw together. Surface lines would have meant the demolition, not of slum housing but of prime commercial properties. The railways therefore had to go underground. In 1886 the North British, and ten years later the Caledonian, opened very costly new east-west lines with lengthy underground sections. These were linked to existing and newly constructed suburban lines. South of the Clyde also new suburban railways were built. For example the Cathcart District Railway, a Caledonian protégé completed in 1894, was planned as a loop to serve new housing developments on the south side of the city. Another late nineteenth-century development, this time independent of the main-line railway companies, was the circular underground line of the Glasgow District Subway Company which opened in 1896.

The problems in Edinburgh were perhaps less pressing, but the same three elements, improvements in station facilities, access to the dock system and growth of suburban traffic were present in most developments. Princes Street station was rebuilt and enlarged by the Caledonian Railway in the 1890s. Waverley posed a bigger problem and was totally rebuilt by the North British between 1892 and 1900. The first line with access to the docks was the early Edinburgh, Leith & Newhaven Railway, later renamed the Edinburgh, Leith & Granton. A second rail link, promoted by the Caledonian, reached Granton in 1861 and Leith western docks in 1864. When a new dock was opened, east of the existing dock system, the Caledonian opened a new branch in 1903. As in Glasgow, the movement of the better-off citizen away from the city centre encouraged the building of commuter lines and the existence of these new railways further encouraged the population to move out. The Edinburgh Suburban & Southside Junction Railway was opened in 1884. Although a suburban passenger line, it had an additional important role as a bypass

Caledonian Railway suburban train at Princes Street station, Edinburgh, about 1900. The short coach, probably a four-wheeler, suggests that this is a train from the Balerno branch. Only four-wheel coaches could be used on the tight curves of this line. SEA

enabling goods trains to avoid the increasingly congested Waverley Station.

The introduction of railways to the lightly populated areas of the western and northern Highlands was a lengthy business. It was characteristic of such lines that they were built in stages, with pauses while money was raised for the next section. However desirable it might have been to build railways in these areas, it was all too obvious to most potential investors that traffic would be sparse and profits doubtful. They were naturally cautious, therefore. The Oban line illustrates the nature of the problem. Callander had a branch line from the Scottish Central Railway by 1858. From there, it took 22 years to reach Oban. The Callander & Oban Railway was opened in 1870 to a station named Killin, which was three miles from the village. Tyndrum was reached in 1873, Dalmally in 1877 and finally, Oban in 1880. A branch to the village of Killin was opened in 1886. The original Killin station was renamed Glenoglehead and a new station built at the junction, appropriately named Killin Junction. Then in 1903 a long branch from Connel Ferry to Ballachulish was completed. North of Inverness, the line to Wick and Thurso was completed in 1874. The Dingwall & Skye Railway, intended to have its terminus at Kyle of Lochalsh, ran out of money and stuck at Strome Ferry in 1870. Kyle was not reached until 1897. The West Highland reached Fort William without needing pause for breath. Construction started in 1889 and the railway was opened in 1894. The Mallaig extension was completed in 1901. Two years later the Invergarry & Fort Augustus Railway, a branch from Spean Bridge on the West Highland, was opened.

There were other proposals for railways in the Highlands, including lines to Ullapool and Lochinver. In 1896 the Light Railways Act had been passed with the intention of encouraging the development of cheap railways in rural areas. The speed of trains was restricted thus allowing the use of lighter track, signalling was simplified and requirements for the fencing of the line were relaxed. The Highland Railway put forward proposals for

250 miles of new railway, including routes on Skye and Lewis, but of these only the Dornoch Light Railway and the Wick & Lybster line were built; they were opened in 1902 and 1903 respectively. Several light railways were also built in the Lowlands, such as that opened in 1901 to Lauder, from Fountainhall on the Waverley route. An interesting line was the isolated Campbeltown & Machrihanish Light Railway, across the Kintyre peninsula. Unlike those already mentioned, this was a narrow-gauge line with only 2ft 3in (686mm) between the rails instead of the standard 4ft 8½in (1,435mm).

Towards the end of World War I, the Secretary of State for Scotland appointed a committee to examine the transport needs of rural areas. In addition to possible new railways it considered roads and steamer services. Most of the railways suggested by the committee were revivals of those proposed by the Highland Railway in the 1890s. None were built, which is just as well. The era of new railway building in Scotland was past and it was already apparent that some of the lines opened around the turn of the century would have been better left unbuilt.

3 Promotion

There were various reasons for the promotion of a new railway. In the early days, the promoters were usually those who expected to profit directly, either because it would be of benefit to other enterprises, such as mines, iron or chemical works in which they had an interest, or it would increase the value of land which they owned along the line of the railway. The long-distance main lines required more capital than could be raised from local landowners and industrialists, and they had to be promoted as good investments for those with money to spare. This money might have originated in land, trade or industry but it would be invested in projects which were seen as giving the best return, rather than in local enterprises. Once the main network was established the initiative for the development of branch lines often came from

Fountainhall station, on the Edinburgh – Hawick line of the North British Railway, was opened in 1848. The route was extended to Carlisle by 1862. Fountainhall became a junction in 1901 when the Lauder branch was opened. SEA

within the community to be served, because of the perceived benefits of a railway connection. There was usually some encouragement from one of the major companies, which would agree to work the new line and in due course more often than not took it over completely. Where new railways were promoted by existing companies there was a strong element of inter-company power politics, not to say megalomania, involved. New railways in remote areas such as the Scottish Highlands were a special case. Because rural depopulation was seen as undesirable, there was considerable government interest in railway development and even a measure of subsidy.

The eighteenth-century colliery waggonway was essentially a private affair. All the ground over which the line was to be built usually belonged to an individual landowner, and all the traffic was from his pits. If matters were a little more complicated and it was desired to cross ground belonging to someone else this could, if both parties were agreeable, be settled by the negotiation of a suitable fee for the right of passage. Later railways were longer, crossing the land belonging to many proprietors and they were providers of public transport available for use by all who wanted to travel or send goods over them. To buy the land and build the lines it was necessary to raise money from a large number of people by selling shares in the enterprise. For such railways the promoters had to obtain an Act of Parliament, authorizing the formation of a company to build and work the line.

The Act was required for two main reasons. First of all the present mechanism for forming a modern-style 'limited liability' company did not exist. The usual way of setting up a business of any sort was by the formation of a partnership, or 'co-partnery', the terms of which were set out in a legal document. The problem with this was that if the business failed, the partners were liable for its debts to an unlimited extent and could lose everything, not just the money which they had put into the business, but all they possessed. One had to be very sure of one's fellow partners in those circumstances! It was not practicable to set up on this basis a business such as a railway company, requiring a large number of people to provide the capital. Few people would be willing to invest. Parliament was deeply suspicious of limited liability companies, thinking that they were too easy to use for fraudulent purposes, and it required therefore a special Act of Parliament each time one was set up. Secondly, the Act conferred powers for the compulsory purchase of the land necessary for the railway. Needless to say there were usually objections from various interest groups and individuals to the granting of an Act. Among the most regular objectors were existing turnpike trusts (responsible for many of the main roads), canal and railway companies which saw the new railway as a threat to their revenue, and landowners who regarded it as a visual intrusion. Often, too, the promoters of new railways were anxious to have power to run trains over parts of existing lines. This might be achieved by negotiation but usually it was incorporated into the Act, giving a further reason for objection by existing railways.

The first railway in Scotland built under an Act of Parliament was the Kilmarnock & Troon. This was a developed version of

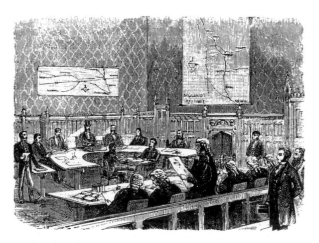

The Parliamentary Committee-room. Before consideration by Parliament a proposal for a new railway went before a Committee of Members. Some of the company directors and senior officers would be in attendance to answer questions.

the colliery waggonway, promoted by a group of landowners led by the Duke of Portland. The Act was obtained in 1808 and the line opened in 1812. Cast-iron plate rails were used, that is rails with flanges on them, instead of the more usual practice of having flanges on the wheels. The main traffic was coal but passengers were also carried. Horses were used for haulage but about 1816 one of George Stephenson's steam locomotives was brought into use, with limited success.

The Monkland & Kirkintilloch, and the Garnkirk & Glasgow were examples of locally-promoted railways. The former was built as a horse-worked line for the transport of coal from the vicinity of Airdrie to Kirkintilloch, for shipment to Glasgow via the Forth & Clyde Canal, in competition with the Monkland Canal. The preliminary meeting held in 1823 to discuss the project was attended by one major local landowner, George More Nisbett of Cairnhill, and by representatives of a second, James Hamilton Colt of Gartsherrie. James Merry of Cairnhill colliery and Walter

Logan, the manager of the Forth & Clyde Canal, were also there. William Dixon of Calder ironworks also became involved at a very early stage. As the industry of the Monklands developed, the main traffic became coal, ironstone and limestone from local mines to the growing number of ironworks near Coatbridge.

The Garnkirk & Glasgow Railway is historically important because it was worked by steam locomotives from the beginning and also because it provided an all-rail route to Glasgow and was, unlike the Monkland, not merely a feeder to a canal. Again the coal traffic was the main thing but a regular passenger service was also operated. When construction started in 1827 the directors included several people connected with the Monkland Railway, including George More Nisbett, James Merry and William Dixon. In addition there were four other landowners and Charles Tennant of St Rollox chemical works. Tennant was one of the major characters in the early industrial history of Scotland. Born in 1768, son of an Ayrshire farmer, he became a weaver at Kilbarchan, Renfrewshire, and then moved into textile bleaching. He developed a bleaching powder which greatly speeded up the bleaching of cloth and set up the St Rollox works to manufacture this new material. By 1825 it had become the largest chemical factory in Europe. Coal was essential for the process and the terminus of the Garnkirk & Glasgow Railway was to be immediately outside the works. It is no wonder that Tennant had a serious interest in the new railway.

With a length of 45 miles, the Edinburgh & Glasgow Railway was a much bigger project than those described above. The Monkland & Kirkintilloch was ten and a half miles long, the Garnkirk & Glasgow eight miles. Reflecting the fact that the Edinburgh & Glasgow was an inter-city main line with easy gradients and gentle curves, the construction cost of £26,000 per mile was more than double that of the Garnkirk & Glasgow and seven times that of the Monkland. This could not be funded in the same way as the others. The Chairman, John Leadbetter and Depute, John Learmonth, were prominent in the commercial circles of Glasgow and Edinburgh respectively but they had to promote the railway far and wide as a good investment in order to raise the money. Over half came from England, and especially from the Liverpool area, where those who had done very well out of the Liverpool & Manchester Railway after it opened in 1830 were looking for similar returns elsewhere.

The North British Railway was first promoted in 1842 as a line from Edinburgh to Dunbar. It proved impossible to raise the money but on the advice of those trying to promote railways from the south, a revised prospectus was issued in the following year for a line as far as Berwick-on-Tweed. Even then money was difficult. Many new lines were being floated at this time and speculation in railway shares came to be seen as a quick and easy way to wealth. This was the so called 'railway mania' and as with all such madnesses, there was the inevitable collapse and financial ruin for many people. In the end the necessary money was gathered, although not without the help of the biggest speculator of them all, George Hudson, who promised that one of his companies, the York & North Midland, would take £50,000 of North British shares. Initially the total English shareholding amounted to 50% and by 1849, when the railway had been open for three years, two-thirds of the shares were held south of the Border. Although the North British eventually became part of an Anglo-Scottish trunk route, it was a route cobbled together piecemeal, rather than planned as a whole. On the other hand, the story of the western Anglo-Scottish route is one of far-sighted planning, inter-company politics, intrigue and government intervention in free-enterprise railway development. One of the key English railways of this period was the Grand Junction. Opened in 1837, this was designed to link the Liverpool & Manchester, opened in 1830, and the London & Birmingham, completed in 1838. In 1836, even before the Grand Junction was finished, the directors had commissioned their engineer, Joseph Locke, to make a preliminary survey for a route northwards to Carlisle and thence to Glasgow and Edinburgh. Once over the border, he looked at the line of the coach road by way of Annandale and Beattock. Finding that the gradients involved would be steeper than he thought prudent, Locke then had a look at a possible route further west, via Dumfries and Nithsdale. By this time efforts were already being made to establish a line from Glasgow to Paisley and Ayr, with a branch to Kilmarnock. The Glasgow-based promoters of this were keen to support the Nithsdale route and only too willing to extend their Kilmarnock branch to meet a line which might be built from the south. When he learned of Locke's views on the route, J J Hope Johnstone, MP for Dumfries-shire, proprietor of the Annandale estates and the largest land-owner in the area, began to pull strings on the Grand Junction board. Locke was told to have a second look and conceded that the Annandale route would be feasible. By this time, the renowned Scottish engineer John Miller, who was the engineer for the Glasgow, Paisley, Kilmarnock & Ayr line, had carried out a survey of the Nithsdale route. With the prospect of two factions applying for Acts to build the Scottish section of the western trunk railway by different routes, the government set up in 1839 a Royal Commission to investigate and advise. In nineteenth-century railway history, such government concern with the choice of route is rare if not unique. Its recommendation came in 1841, to the effect that only one cross-border route was ever likely to be necessary to serve both Glasgow and Edinburgh, and that being so the Annandale route was preferred. This advice was soon overtaken by events and by 1850 there were three routes in operation, the Caledonian via Annandale, the Glasgow & South Western by Nithsdale and in the east, the North British via Berwick. All three are still in operation today.

By 1850 most of the important main-line railways in central Scotland were in operation or under construction. There then followed a spate of branch lines to provide links to the main stem. Generally these were promoted by local people, with the promise of some measure of assistance from the company whose line the branch joined. While some tried to run their own trains for a time,

Pass issued to a director of the Dundee & Arbroath Railway. This line was opened in stages between 1838 and 1840.

many were worked from the start by the main-line company. The financial arrangements varied but it was generally on the basis that the operating company received a certain percentage of the gross takings. Out of the remainder, the owning company had to maintain the line and, if there was anything left, pay a dividend. Either way, after a period of independence, there was usually a complete take-over.

The St Andrews Railway was a typical example of this process. One of the lines of the Edinburgh, Perth & Dundee Railway (originally called the Edinburgh & Northern) ran from Burntisland to Tayport, via Ladybank, Cupar and Leuchars. It was opened to Cupar in 1847 and Tayport in 1850. A group of people in St Andrews, including Provost Playfair and Robert Haig of the whisky-distilling family, then promoted the St Andrews Railway to build a four-and-a-half-mile line from the town to the main line. The necessary Act was passed in 1851. From the start the venture was supported by the Edinburgh, Perth & Dundee. They built a new station at the junction and agreed to supply the necessary locomotive, carriages and waggons, and work the line. Technical advice from their engineer, one Thomas Bouch, was made available to the new company. Before long Bouch left to set up on his own as a consulting engineer and the St Andrews Railway became his first client. The new railway opened for business on 1 July 1852. At first it paid its way, but with the passage of time and growth of traffic the maintenance of the track and of two wooden bridges became an increasing burden. However, the St Andrews Railway Company struggled on and kept its independence until 1877, when it was taken over by the North British

Railway which had already absorbed the Edinburgh, Perth & Dundee fifteen years earlier.

Local initiative proved to be as essential for the development of major routes in remote areas as it was for branch lines such as the St Andrews Railway. There had been talk of a railway to open up the West Highlands since the 1840s. In the early 1880s the Glasgow & North Western Railway was proposed. This would have been 167 miles long, stretching from Glasgow to Fort William and then up the Great Glen to Inverness. There was massive opposition from the Caledonian Railway, the Highland Railway, steamboat operators and landowners. As a result, the necessary Act was not granted. However, the climate of opinion was changing and the need for improved transport in the Highlands was becoming more generally accepted. With the backing of the North British and their promise to work the line, another less ambitious scheme was promoted, this time for a railway to Fort William only. The proposed route went across Rannoch Moor, an area which even now is without roads. Although there was some opposition, the proposal was generally welcomed in Fort William and had the blessing of all five proprietors through whose land the 100-mile line would run. The West Highland Railway was duly authorized in 1889 and opened to Fort William in 1894. It had always been hoped to extend the railway to the west coast. In 1894 the construction of the railway to Mallaig was authorized, but it was realized by all concerned that this could never be viable and the line would have to have a Government subsidy if it was to be built at all. However, there was considerable opposition in Parliament to this and it was not until 1896 that the West Highland Railway

(Guarantee) Act was passed. Under its terms, a dividend of 3% was guaranteed on the Mallaig extension capital of £260,000 and a grant of £30,000 was made towards the £45,000 cost of a pier at Mallaig. The new line was opened in 1901. Both the West Highland line and the Mallaig extension are still in use.

Not many new railways were built in Scotland after the Mallaig line. The Ballachulish branch of the Callander & Oban, and the Invergarry & Fort Augustus Railway were both opened in 1903. Various minor lines, some promoted under the 1896 Light Railways Act, were also completed in the early years of the twentieth century. All were closed many years ago. Several other projects which were enthusiastically promoted were abandoned after they had been authorized, or even after construction had started. The Mallaig line therefore has the distinction of being Scotland's newest railway of any great length which is still in use!

4 Planning

At a very early stage in any new railway scheme the engineers were called in. They carried out preliminary surveys of possible routes and prepared plans and cost estimates which the promoters required for presentation to Parliament. The engineers had to be prepared to attend Parliamentary Committees and be questioned about the proposals. Once the Act of Parliament authorizing construction had been obtained the real work of the engineers could begin. The works had to be designed for economical construction, coupled with satisfactory operation and minimum maintenance costs. Detailed plans and specifications were prepared so that offers for the construction could be sought from contractors. The various offers submitted were then considered and recommendations made to the directors. Once the contracts had been placed and construction started, the engineers were responsible for overseeing the contractors, ensuring that the work was being carried out in accordance with the plans and specifications, and authorizing payments as the work progressed.

Surveyors at work taking ground levels. In the right background there is a man holding a graduated staff. This is being read by the surveyor using a level, and the figures are being recorded by the assistant on his left. A preliminary survey had to be made in order to prepare plans for submission to Parliament. If construction of the line was authorised a more detailed survey was made so that contractors could be asked to submit tenders for the construction work.

Where did the engineers for Scotland's first railways come from? Some came from a background in mining; others had worked on canals and roads. Because the development of railways in England began earlier than in Scotland, some engineers who had gained their experience south of the Border worked on Scottish lines. Some of the well-known English engineers, for example George Stephenson, became involved in Scottish railway building, but the extent of this was usually limited. They were called in as arbiters, as 'names' to add credibility to a prospectus or reassure the English shareholders who provided a large part of the money for some lines. Only one or two English engineers made any sustained contribution to the development of the Scottish network. Very soon, within Scotland there were many engineers with the expertise necessary to cope with the largest projects.

The most important railway engineers in Scotland were Thomas Grainger and John Miller. Although now little known, separately and in partnership they were connected with most of the Scottish railways built in the second quarter of the nineteenth century, and they were also involved in developments south of the Border. Thomas Grainger was born in 1794 at his father's farm, Gogar Green, just west of Edinburgh. At the age of sixteen he started work with the Edinburgh engineer, John Leslie. By 1816 he felt able to set up on his own as a civil engineer and surveyor, engaging mainly in road works. Grainger's future partner, John Miller, was born in Ayr in 1805. He went to school in Ayr and then attended Edinburgh University. Back in Ayr, he began to serve a law apprenticeship but changed direction and decided to become an engineer. In 1823 he became an assistant to Grainger and two years later he became a full partner.

The Monkland & Kirkintilloch Railway, opened in 1826, was probably their introduction to railway engineering. Promotion of this had begun in 1823 and Grainger had been involved at an early stage. The partnership then worked on the Edinburgh & Dalkeith and the Garnkirk & Glasgow Railways, both opened in 1831. When more ambitious projects were being considered a few years later, it was frequently to Grainger and Miller that the promoters turned. They were engineers for the Glasgow, Paisley, Kilmarnock & Ayr and the Edinburgh & Glasgow Railways. A route for a railway between Edinburgh and Glasgow had been surveyed by James Jardine in 1825 but there was no immediate action. Five years later there was revived interest and Grainger and Miller undertook a survey. The advice of George Stephenson was sought and he agreed with the route suggested by the engineers but the first attempt to obtain an enabling Act failed. However, it was inevitable that a railway would be built sooner or later between Scotland's two most important cities and the necessary Act of Parliament was eventually obtained in 1838 and the line opened in 1842. The railway was largely the work of John Miller, no doubt because Grainger was fully occupied with other projects. Over half of the money came from England, and in order to reassure these shareholders Miller was required from time to time

Forming an embankment. Material is being carted along a temporary railway and tipped over the end of the lengthening embankment. In planning a new railway the engineers always tried to arrange things so that the volume of material required to create embankments matched that which had to be excavated to form cuttings.

to consult the prominent English engineers, Joseph Locke, John Rastrick and Charles Vignoles.

Sometime in the mid-1840s the partnership was dissolved, Miller taking a large part of the business with him. The writer of an obituary some 40 years later said 'His practice during the period of the great railway mania was probably as extensive as that of any other engineer in the United Kingdom. In the year 1845 he deposited plans for between 1,500 and 1,600 miles of railway. He never promoted any railway, but simply worked for those capitalists who required his professional assistance, and his services as a witness before Parliamentary Committees for the promotion of railways.' For any good engineer who could stand the pace, the period of the railway mania was a golden opportunity. Of course he could not do everything himself and he had many pupils and assistants, some of whom subsequently had successful and distinguished careers. By 1849 Miller had made his fortune and decided to retire. There is little information on what he did in his long retirement except that he was a Member of Parliament from 1868 to 1874. He died in Edinburgh in 1883. Sadly, his former partner Thomas Grainger did not live to retire. He died as the result of an accident on the Leeds Northern Railway in 1852, at the age of 57.

Of all the British railway engineers of the nineteenth century, the partnership of Joseph Locke and John Errington was remarkable for the quantity and quality of work produced. They were among the very small number of English engineers to be responsible for major projects in Scotland. Although they did a lot of work in Scotland this formed only a small proportion of the total mileage they undertook both in Britain and abroad. Locke came from a mining background. He was born near Sheffield in 1805, the fourth and youngest son of a colliery manager. He became a pupil of William Stobart of Pelaw, Durham, who was a 'colliery viewer', or mining engineer. Then in 1823 he went to George Stephenson, with whom he later worked on the Liverpool & Manchester Railway. In 1832 he had a difference of opinion with Stephenson over errors discovered in part of the survey. As a result of this Locke left to start on his own as a railway engineer.

Errington came from Hull and was born in 1806. His early experience was on various works in Ireland. He then returned to England and was soon engaged in railway surveying. His association with Locke started while they were both working on the Grand Junction Railway. Although between them Locke and Errington undertook a vast amount of work at home and abroad, in the Scottish context the important project was the Grand Junction. This was seen as a key link in a great north-south trunk route stretching into Scotland. As already mentioned, Locke was sent on an exploratory mission to look for possible routes as far as Glasgow and Edinburgh, and beyond. The result was a series of railways, promoted by companies which were at least nominally independent but conceived as an entity, all engineered by Locke and Errington, and forming the successive parts of a great main line, the present-day West Coast Main Line.

The most southerly Scottish section was formed by the Caledonian Railway, from Carlisle to Glasgow, Edinburgh and Castlecary. The route was continued northwards to Aberdeen by the Scottish Central, Scottish Midland Junction and Aberdeen Railways, together with several short lines already in existence. Locke and Errington were concerned, too, in the earlier Glasgow, Paisley & Greenock. On top of all his railway work Locke became a Member of Parliament, for Honiton, Devon. He died in 1860 at the age of 55. Errington also died aged 55, in 1862. The life of the nineteenth-century railway

Joseph Mitchell, 1803-83, engineered many of the railways in the Highlands, including the main line between Perth and Inverness.

engineer was hard and few reached a ripe old age. John Miller, who was fortunate in having both the money and the temperament which enabled him to retire early, was one of the few.

Another survivor was the Highland engineer Joseph Mitchell who reached the age of eighty. Unlike Joseph Locke and John Errington, who were almost exclusively railway engineers but were geographically unconfined, Mitchell did little outside the Scottish Highlands, but was responsible for roads, harbours and even churches, as well as railways. Joseph Mitchell was born at Forres in 1803. His father, John Mitchell, worked for the Commission for Highland Roads and Bridges, under Thomas Telford, and became Chief Inspector. With this background Joseph was destined to be an engineer. On the advice of Telford, who had himself started his career as a mason, Joseph first went to work with the masons building the Caledonian canal locks. In 1821 Telford offered him the chance to work in his London office, a busy, bustling place with a wide range of work passing through it.

John Mitchell died in 1824 and at the age of 21 Joseph was appointed as his successor, a remarkable burden of responsibility for one so young. He continued to work for the Commission for 40 years and in time also began to undertake railway and other work which, if not directly for the Commission, must have had their blessing. His first railway job was a survey in 1838 of an alternative route for the Edinburgh & Glasgow Railway. This was on behalf of the Earl of Hopetoun, because part of the line then being proposed encroached on his estate, but it had no effect on the route finally used. His preliminary survey in 1844 of a route for the Scottish Central Railway also came to nothing, and Locke and Errington became the engineers.

However, Mitchell had acquired a taste for railway engineering and was confident that the knowledge gained in the course of his work on Highland roads could be put to good use. He was the engineer for the direct line from Inverness to Perth over the Drumochter pass. This was first promoted in 1845 but it was 1861 before construction was authorized. The line was opened in 1863. In the interlude, Mitchell had surveyed the routes for other lines in the Highlands, including the Inverness & Nairn, opened in 1854 and extended to Elgin and Keith four years later. He also surveyed parts of the line northwards from Inverness and westwards to Kyle of Lochalsh. Unfortunately he suffered serious illness in 1862 and he made only a slow recovery. Wisely, he then took into partnership two of his assistants, William and Murdoch Paterson. Five years later he retired, and Murdoch Paterson in particular continued with the construction of several of the railways surveyed earlier by Mitchell. Paterson then undertook various other works, including the completion of the line to Wick and Thurso.

Mitchell devoted his retirement to writing his reminiscences which give a lively picture of life in the Highlands in the mid-nineteenth century. He died in 1883, aged 80, after a remarkable career. His contribution to the development of the transport infrastructure of the Highlands was invaluable. He not only acted

The world's first roll on-roll off train ferry, used to take goods waggons across the Forth. This came into operation in 1850. Thomas Bouch designed the ingenious loading ramp, or 'flying bridge', to accommodate tidal changes.

as the engineer but, in addition, the wide range of contacts he had made in the course of his work with the Commission for Roads and Bridges enabled him to become involved in the promotion of much needed railways. Few engineers had the opportunity or the ability to take on this additional role.

Anyone asked to name a nineteenth-century railway engineer who worked in Scotland would be most likely to come up with Thomas Bouch, if they knew of any at all. Responsible for the first Tay Bridge, Bouch is inevitably notorious rather than famous. However, there was more to the man and his work than a bridge which failed. Bouch was born in 1822 at Thursby, near Carlisle, the son of a sea captain. He worked briefly with a mechanical engineering firm in Liverpool in 1840 and then got a job on the construction of the Lancaster & Carlisle Railway. Late in 1844 he went to the Stockton & Darlington Railway. He stayed there for just over four years before becoming manager and engineer with the Edinburgh & Northern Railway early in 1849. Renamed the Edinburgh, Perth & Dundee Railway later that year, this suffered from the disadvantage that most of its mileage was in Fife and it could serve the cities in its title only by courtesy of other railways and by the use of ferries across the Forth and Tay. As manager Bouch was only too well aware of the problem and as engineer he set out to develop a solution. The result was the world's first roll on-roll off train ferries, from Granton to Burntisland and Tayport to Broughty Ferry. The ingenious loading ramps, or 'flying bridges', were designed by Bouch and the ships were built by Robert Napier on the Clyde. The first vessel, *Leviathan*, began operations on the Forth in March 1850. The Tay service started in the spring of the following year with the *Robert Napier*. These were for goods waggons only and passengers still had to change trains and cross by a connecting steamer. However, the train ferries greatly facilitated the important goods traffic and they continued to run until the completion of the bridges.

The success of the train ferries enhanced Bouch's reputation as an engineer, and perhaps too, he felt more at home in engineering and than in management. He resigned in 1851, soon after the ferry service was fully operational and set himself up in Edinburgh as a consulting engineer. The time was ripe, as there

The first Tay Bridge, built by Thomas Bouch and destroyed in a gale.

were many new railways in the offing. A considerable number of these were small-scale local affairs intended to provide links with the established main lines. They were promoted by local people with no knowledge of railways and little money. A substantial part of Bouch's engineering practice in Scotland and the north of England was aimed at meeting this need for cheap local railways. Among them were several in Fife, including the St Andrews, Leslie, and Leven lines. In the Edinburgh area he engineered the 21-mile Peebles railway, the Edinburgh, Loanhead & Roslin Railway and the Edinburgh South Side & Suburban Junction Railway. In all he was responsible for more than 120 miles of line in Scotland. There was sometimes a price to be paid. While Bouch's railways were cheap, they were sometimes too cheap to be durable, as the shareholders soon found out.

Bouch's work was not all on small-scale minor railways. Among his north of England projects was the 50-mile South Durham & Lancashire Union Railway across the Pennines. This had two large viaducts built of iron. There had been talk of bridging the Tay and Forth at least from the early 1860s and when a Tay bridge came to be seriously considered towards the end of the

North British Railway no 224, built in 1871, the locomotive which went down with the first Tay Bridge in 1879. The damage was not nearly as bad as it looks! After repair 224 went back into service and ran until 1919, nicknamed 'The Diver'. SEA

decade, Bouch was appointed engineer. The Act authorizing construction was passed early in 1870. Work started in the summer of the following year and the bridge was opened for traffic on 1 June 1878. Queen Victoria crossed it in the summer of 1879 and on 26 June its engineer received a knighthood at Windsor. The collapse of the bridge in a gale six months later is too well known to need much comment here. The Court of Enquiry into the collapse roundly condemned Bouch for defects in the design, construction and maintenance of the bridge. While as the engineer he did carry full responsibility such a blanket condemnation was rather hard. The damning report of the Enquiry was published on 5 July 1880 and it was clear then that whatever the merits of the case, neither Parliament nor public opinion would accept a rebuilt bridge with which Bouch was associated. The report was obviously a great blow to Bouch but he carried on working. On the night of 6-7 August, while returning from a visit to London in connection with the Edinburgh Suburban Railway, he was taken seriously ill. Doctors advised complete rest and Bouch went to Moffat, but he never recovered his health and died there on 30 October 1880.

There were many other notable engineers involved in the creation of Scotland's railway system. Among them was Benjamin Blyth who worked with Grainger and Miller before leaving in 1850 to set up his own firm in Edinburgh. This did good work before and after his death and still exists today. Charles Forman, of Formans and McCall, Glasgow, was the engineer for the West Highland Railway, the Glasgow Central Railway running east to west under the streets of the city and many others. There were, too, the specialists called in for particular jobs, such as the design of large bridges. The second Tay Bridge was the work of W H and Crawford Barlow of London, who made use of some of the

Holiday traffic was an important source of revenue. The railway reached Fraserburgh, on the Great North of Scotland Railway, in 1865. This LNER poster dates from the 1930s.

H. G. Gawthorn.

FRASERBURGH
GLORIOUS SANDS · BATHING · GOLF · TENNIS

BOOKLET FROM TOWN CLERK (DEPT P) OR ANY
L · N · E · R INQUIRY OFFICE

Published by THE LONDON & NORTH EASTERN RAILWAY   PRINTED IN ENGLAND   THE DANGERFIELD PRINTING CO LTD LONDON

One of the most imposing railway structures in the Highlands, the Glenfinnan viaduct on the Mallaig extension of the West Highland Railway is still very much in use. A notable feature of the structure is the use made of mass concrete rather than masonry. Ian Larner

The signalbox at Glenfinnan. Radio-signalling is now used to control the running of trains. Ian Larner

The approaching steam locomotive is an LNER design, probably one of those displaced from the East Coast Main Line by diesel locomotives. The photograph was taken between Larbert and Stirling in the early 1960s. J L Wood

Caledonian driver John Souter of Perth and his locomotive, no 17. On the last night of the 1895 'Railway Race' between the companies operating the competing east-coast and west-coast routes from London to Aberdeen, Souter took the west-coast train from Perth to Aberdeen at a record speed of 66.8 miles/hr (107.5km/hr). It is likely that the carriages in this picture were drawn in, as they are out of scale with the engine.

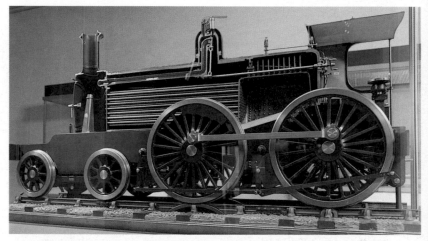

Model of the North British Railway express locomotive Abbotsford, *one of twelve built in 1876 and 1878. The design, by Dugald Drummond the company's locomotive superintendent, had a great influence on British locomotive development for several decades. The model was made in the Museum workshop using drawings supplied by the North British Railway.*

Model of the North British Railway express locomotive Waverley. *A total of 22 locomotives of this type were built between 1906 and 1921. The model is finished in the colours of the LNER, as used after the grouping of the railways in 1923.*

Opposite: *Signalbox lever frames are used to control the signals and points. This frame was originally at Fountainhall Junction but was moved to Edinburgh by British Rail and used to train maintenance staff. See page 22 for a photograph of the station and signal box at Fountainhall.*

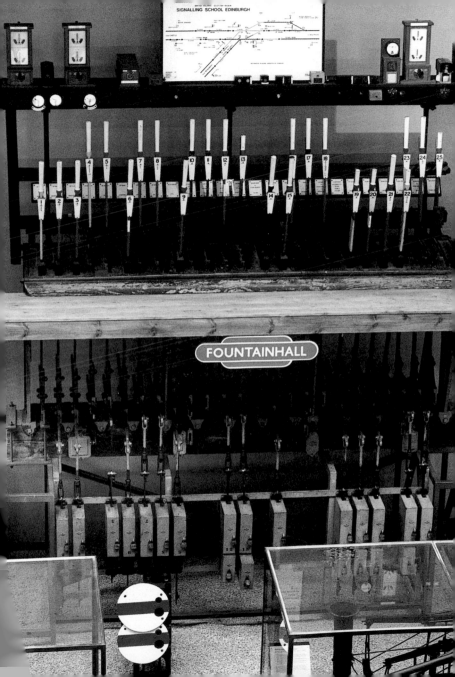

SIGNALLING SCHOOL EDINBURGH

FOUNTAINHALL

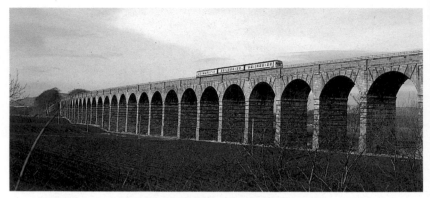

The Almond viaduct on the Edinburgh & Glasgow Railway, completed in 1842 and still very heavily used.

Ellesmere *is the oldest surviving Scottish-built locomotive in Britain. It was built in 1861 by the Leith firm of Hawthorn and worked at Howe Bridge colliery, near Manchester, until 1957. Many collieries, ironworks and other industrial sites had their own locomotives for internal use.*

The Slamannan basin at Causewayend, on the Edinburgh &
Glasgow Union Canal. This was the original terminus of the
Slamannan Railway, opened in 1840. The line was later
extended to Boness. Angus and Patricia Macdonald

George Graham was the Caledonian Railway's chief engineer from 1853 to his death in 1899.

girders from the earlier structure, and the Forth Bridge was designed by John Fowler and Benjamin Baker, also of London.

The engineers so far mentioned were consulting engineers, retained by the promoters of new lines or by established railway companies to undertake particular pieces of work. However, the companies themselves had to set up engineering departments to deal with maintenance work and as these grew in size and experience they became capable of tackling many of the new works, although consultants were also retained as required. One of the most notable 'in-house' engineers was George Graham of the Caledonian Railway. He was born in Dumfries-shire in 1822. In 1845 he was working with Locke on the survey of the Caledonian. He joined the staff of the railway company and in 1853 became chief engineer responsible for both maintenance and new works. He remained in the job until his death in 1899 at the age of 76, although latterly he was in charge of new works only.

5 Building the railways

Successful completion of a major railway project meant glory for the engineers and sometimes substantial profits for the contractor. Railway contracting was an uncertain business and the failure rate among contractors was high. Often losses occurred through no fault of the contractor; there were technical difficulties which

Forth Bridge, designed by Benjamin Baker and John Fowler and completed in 1890. This was the first major structure to be built of steel, rather than wrought iron. Ian Larner

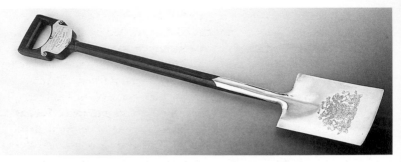

*Ceremonial spade used to cut the first sod for the Berwickshire
Railway in 1862. Most of it was opened in November 1863 but
the final link, the nineteen-arch Leaderfoot viaduct, was not
ready for another two years.*

could not be discovered until work started, or there was a scarcity
of labour which pushed up wage rates.

The way in which railway construction was organized is best
illustrated by taking as an example the 27-mile branch from Connel
Ferry, on the Callander and Oban line, to Ballachulish. This was
one of the last sizeable railways to be built in Scotland; work started
in 1898 and the line was opened in 1903. Duncan Kennedy, who
worked on the railway at the start of his career as a civil engineer,
gives a fascinating account of its construction in his book *The Birth
and Death of a Highland Railway* (1971). The contract was awarded
to a Glasgow firm, John Best. In charge of the whole of the works
on site was the contractor's 'agent', who in this case was John Best's
son, Allan Best. Under him was a chief engineer, a Mr Wilson, and
three section engineers, who were responsible for ensuring that the
correct line was followed, that the track levels were correct and that
generally the work was in accordance with the drawings and specification. There was also a cashier and each section had a time-
keeper. The men worked in gangs, under a 'ganger' (foreman) and
the overall supervision of labour was the responsibility of Alick
Addison, the 'walking ganger', so called because he was constantly
on the move over the full length of the contract.

In parallel with the contractor's organization there was a team from the engineers, Formans and McCall, led by the 'resident engineer', A J Pringle, aided by assistant engineers and inspectors for each section. Formans and McCall had designed the new railway for their client, the Callander & Oban Railway. Their resident engineer was therefore the guardian of the client's interests, and he and his staff had to ensure that the work was carried out as specified. The contractor was paid monthly and one of the most important duties of the resident engineer's team was the measurement of the work done each month to determine the payment due.

In the early days, however, there were no large contracting firms capable of building complete railways in this way. The engineers therefore had to organize the construction as well as undertaking the design. This might involve the railway company employing labour directly and buying all materials, equipment and tools. More usually, the work was let out in a large number of small contracts, according to the capacity of the local contractors available, with the engineers retaining overall responsibility for co-ordinating their activities. Most of these contractors are now virtually unknown. For the Garnkirk & Glasgow Railway,

Navvy and ganger (foreman) on the Ballachulish branch of the Callander and Oban line, completed in 1903. The drawings are by D T Rose, one of the assistant engineers supervising the construction of the railway.

engineered by Grainger and Miller, the earthwork, that is the formation of the cuttings and embankments necessary to provide an easily graded bed for the track, was split into two contracts. This was for a line with a total length of only eight miles. Here the names of the contractors have survived. They were Forbes and Sutherland of Edinburgh for the eastern section, and Riddell and Thomson for the western. It is recorded that the directors also contracted directly with McCulloch and Company, Glasgow, for cast-iron chairs (on which the rails sat) and with a Mr Keith of Dundee for sleepers. In the case of the 47-mile Edinburgh & Glasgow Railway, for which Grainger and Miller were also the engineers, there were 21 contracts including two separate ones for large viaducts. The first contracts were let in 1838 and within a short time there were 7,000 men and 700 horses at work on the line. Among notable contractors involved in later railway building in Scotland were the London firm of Lucas and Aird, who built the West Highland Railway, and Robert McAlpine and Sons, Glasgow, builders of the West Highland extension to Mallaig.

In a different class altogether was the huge firm of Thomas Brassey (1805-1870), whose operations and reputation were international. Brassey built many miles of railway in Scotland. In addition he worked in England, Wales, France, Belgium, Holland, Spain, Canada, India and Australia, among other places. He won his first railway contract in 1835, for a ten-mile stretch of the Grand Junction Railway. The engineer of the Grand Junction was Joseph Locke, who became engineer for the Glasgow, Paisley & Greenock Railway. This was built by Brassey between 1838 and 1841. The combination of Locke and Brassey went on to greater things when, as mentioned above, Locke became engineer for the various companies which continued the west-coast line right

Rock cutting at Bishopton on the Glasgow, Paisley & Greenock Railway. The railway was completed as far as Paisley by 1839 but because of the difficult work involved in the cuttings and tunnel at Bishopton it was 1841 before the line was completed throughout.

through to Glasgow, Edinburgh and Aberdeen. Work on the Caledonian Railway, the southernmost Scottish link in the chain, was well under way by the middle of 1846. In the following year there were 20,000 men on the job. The route was opened for traffic to Glasgow and Edinburgh in 1848 and to Aberdeen in 1850. Brassey had other contracts in progress at the same time and in the late 1840s he was employing some 75,000 men. His connection with Scotland did not end with the completion of the railway to Aberdeen; after 1850 he was involved in the building of other lines, including the Inverness & Nairn, opened in 1855.

Brassey's agent on many of his Scottish contracts, including the Scottish Central and Scottish Midland Junction Railways, was James Falshaw who was born in Leeds in 1810. He was one of the many less well-known railway engineers who, although obscured by the 'great men', were indispensable to the successful completion of their works. Originally trained as an architect and surveyor, he first became involved in docks and waterworks. After working for Brassey, he must have decided that he liked living and working in Scotland because he made his home in Edinburgh. In 1861 he became a director of the Scottish Central Railway. In 1881 he was deputy chairman of the North British and Chairman from 1882 to 1887. Railway contracting continued for a time, however. Independently of Brassey, he built the Berwickshire Railway in 1862 and his final job a few years later was a branch line in the north of England, for the North Eastern Railway. He died full of honours in 1889, at the age of 79, as Sir James Falshaw, JP, DL, FRSE.

While a single contractor was usually given complete responsibility for a substantial length of line, it was common practice for major bridges to be the subjects of separate contracts with specialist firms. On the Ballachulish branch for example the main contract went to John Best but the two major steel bridges, at Connel and Creagan, were built by Arrol's Bridge and Roof Company, Glasgow. This firm is not to be confused with that of William Arrol and Company, also of Glasgow, builders of the second Tay

Bridge and the Forth Bridge, completed in 1887 and 1890. William Arrol (1839-1913) was born at Houston, Renfrewshire and trained as a blacksmith in Paisley. He established his own business in 1868 and built it up to become the leading firm of bridge builders, and structural and hydraulic engineers, working both in Britain and overseas. He was knighted following the official opening of the Forth Bridge by the Prince of Wales.

The tale of woe concerning the collapse of the first Tay Bridge in a gale on 28 December 1879, nineteen months after its opening, is well known. Although the Court of Enquiry blamed the engineer, Thomas Bouch, he was plagued by problems with the contractors and these certainly contributed to the disaster. The contract went in 1871 to Charles de Bergue of London. De Bergue himself died in 1873 and although the firm struggled on until the following year, in the end they had to give up. The contract was then taken over by the Middlesborough firm of Hopkins, Gilkes and Company. The price de Bergue had quoted was far too low. In an effort to cut costs, a makeshift foundry had been set up on the south bank of the Tay at Wormit, to produce the many cast-iron columns required. Unfortunately, the quality of the castings was poor. If instead of setting up the Wormit foundry, the contractor had gone to one of the established foundries in Dundee, it is unlikely that the bridge would have failed. Nevertheless Bouch was undoubtedly at fault in not arranging for adequate inspection of the foundry work and indeed other aspects of the construction of the bridge.

There is usually some information available about the engineers and the larger contractors. In contrast, the anonymous toiling thousands who actually built the railways by their strength and skill have left little trace as individuals. Who were these 75,000 men who formed Thomas Brassey's army engaged on works for which he had contracted in the late 1840s? Where had they come from? What was their impact on the countryside in which they worked?

The last question is perhaps the easiest to answer; it was rather like that of any invading army, causing consternation and alarm among local residents. The author Thomas Carlyle, who was born in Ecclefechan, wrote in August 1846 to a friend:

> The country is greatly in a state of derangement, the harvest with its black potato fields, no great things, and all roads and lanes overrun with drunken navvies; for our great Caledonian Railway passes in this direction, and all the world here, as everywhere, calculates on getting to heaven by steam! I have not in my travels seen anything uglier than that disorganic mass of labourers, sunk three-fold deeper in brutality by the three-fold wages they are getting. The Yorkshire and Lancashire men, I hear, are reckoned the worst; and not without glad surprise I find the Irish are the best in point of behaviour. The postmaster tells me several of the poor Irish do regularly apply to him for money drafts, and send their earnings home. The English who eat twice as much beef, consume the residue in whisky, and do not trouble the postmaster.

In addition to the English and Irish mentioned by Carlyle there were many Scots engaged on railway works. The Highland Clearances and the Irish famine both drove people to seek a livelihood elsewhere and many became navvies. The term navvy is a relic of the canal era of the eighteenth and early nineteenth centuries, when the men who excavated these artificial waterways became known as 'navigators'. The first railway navvies would very likely have worked on canals, or possibly on road building. Although some labour for railway building was recruited locally for each project (it usually paid much better than farm work), by the time the Caledonian Railway was being built the railway navvies had emerged as a distinct group. Navvies lived together in encampments on the job, and tramped from contract to contract, following the work. They were engaged in the hardest physical labour, excavating, tunnelling, bridge building, and they ate and drank prodigiously. An agricultural labourer might become a navvy. Starting on a local job, becoming fitter and tougher in the process provided he could stand the work, he might then take up the peripatetic way of life of the navvy.

The work was dangerous and living conditions were, in general, dreadful. In the early days the navvies threw together their own accommodation using whatever material they could find on site. In later years it became the custom for the contractor to build hutted camps. Although the pay was good there were many ways of parting the navvy from his hard-earned money. Often living and working in remote areas, far from any town or even village, he was dependent for the necessities of life on the shop established on the site. His employer possibly had a financial interest in this. Pay days were infrequent, probably only monthly, and to enable a man to survive between pays he could apply for a 'sub' or subsistence allowance. This would be not in cash, but in the form of a ticket which could be spent only in the shop. These shops, known as truck shops or tommy shops, were notorious for high prices, poor quality (tommy rot) and short weight, and also for pushing the commodity which paid them best, drink.

Another hazard was the dubious sub-contractor, although the better contractors were very particular in choosing sub-contractors. It was rare for the main contractor to employ all the men working on a job directly. Instead, parts of the contract would be sub-contracted, and sometimes several layers of sub-contractor might be involved. Sub-contractors as a group had rather a doubtful reputation and while most were honest there were the fly-by-nights who would decamp with the money just before pay day.

It is not surprising therefore that the navvies worked hard and played hard. If despite the subs and the tommy shop there was any money left on pay day they were liable to go on a 'randy', drinking all they had. There was little or no work done for several days afterwards, and this was sometimes put forward by contractors as a reason for paying the men infrequently. Of course after the randy they had no money and application had to be made for a sub, which could only be spent in the tommy shop and so the cycle continued. In the mid-1840s masons working on the construction of the North British Railway who were earning 4/- (20p) a day, were receiving cash wages of 10/- (50p) a month, or less, because of this system.

Work in progress on a cutting, believed to be on the Newburgh & North Fife Railway. This was opened as late as 1909. SEA

Railway navvies as a group acquired a reputation for drunkenness and fighting, often with religious or ethnic differences as the excuse, but individually they are anonymous. On occasion the fighting became rather more serious, involving major riots and murder, and a few individuals acquired a transient sort of fame as a result of their behaviour. The worst incident of this kind in Scotland took place in 1846. A group of Irishmen working on the North British Railway's Edinburgh to Hawick line became involved in a disturbance in the pub in which they had just been paid, in Gorebridge, Midlothian. Two of them were arrested but 200 equipped themselves with pickaxes and other weapons and freed the captives from the local police station. Afterwards they chanced to meet two policemen, Richard Pace and John Veitch, and beat them so severely that Pace later died. A mob of 1,000 Scots working further south, led by a piper and a bugler, set out on a revenge mission. When they reached the area they were reinforced by 150 men from local collieries. The Irishmen wisely fled when they saw the size of the approaching mob which then proceeded to burn their encampment. By the time a detachment of police from Edinburgh, accompanied by 60 dragoons, reached the scene, the rioters had dispersed. The next day 200 Irishmen set out from Edinburgh to meet the Scots but they were intercepted by the police and soldiers and persuaded to go back to the city peacefully. No arrests were ever made in connection with Pace's murder although two wanted men, Peter Clark and Patrick Reilly, were named and a reward of £50 offered for information as to their whereabouts.

While it may not be acceptable as an excuse for riotous behaviour leading to murder, it can be suggested as a reason, that working and living conditions on this contract were even worse than usual, probably because of the poor state of the finances of the North British Railway. The Irishmen were in an ugly mood because of the conditions and they were also sure they had been swindled by the tommy shop. The Scots further south probably felt much the same.

A few years earlier, in 1841, on the Edinburgh & Glasgow Railway, Dennis Doolan and Patrick Redding were hanged for the murder of their ganger, John Green, in December the previous year. This seems to have been the settling of a personal score, arising out of something which happened when they worked together on a previous job in England. The execution took place at the scene of the murder, Crosshill bridge near Bishopbriggs, with a large military presence because it was feared that there might be a rescue attempt by the mob.

If few navvies are remembered as individuals, and those few mostly for disreputable reasons, their collective monuments are all around in the form of the still operating railways which they built. By the standards of today's creeping motorway programme, even with all the machinery now available, the achievement of the railway navvies of last century with pick, shovel and wheelbarrow are colossal and hard to comprehend.

6 Operation

If building the railways was a major task, in some ways operating them was even more difficult. This was a new and rapidly expanding industry requiring in its staff, old skills adapted to meet new needs, or completely new skills. Administrators and clerks, civil and mechanical engineers, engine drivers and firemen, station staff and signalmen, were all needed in large numbers. For the very first railways there was, of course, no pool of experienced staff from which to draw. Staff had to be recruited from other industries with related skills, for example mechanical engineers with experience of marine or factory steam engines, clerks and commercial staff from other businesses and guards from those made redundant when the railway replaced road coach services. Other skills had to be learned from scratch. There was no outside precedent for train operation, driving and signalling, and building up the necessary pool of experience was sometimes a painful process for both staff and passengers. Inevitably, after the first few railways had become well established, their staff were poached by the newcomers and so it went on as long as the industry expanded.

In addition to the staff employed by the railways themselves, the Government, through the medium of the Board of Trade, required staff to inspect new lines, investigate accidents and generally see that the relevant regulations were as far as possible enforced. The obvious pool of engineering expertise from which to recruit an independent railway inspectorate was that of the army engineering officers. Many of these had distinguished second careers as Board of Trade inspectors.

The organizational structure of the railways, with its strict code of discipline, uniforms and hierarchical grading of staff, had a strong military flavour. The senior officers recruited by the early companies came from various backgrounds but there was a considerable sprinkling of military men. The most influential of these was probably Captain Mark Huish, who began his railway career in

Scotland. He came to railway work in 1837 from the army of the East India Company, at the age of 29, as secretary of the new and not yet open Glasgow, Paisley & Greenock Railway. In 1841 he became general manager of the Grand Junction Railway. By a series of amalgamations masterminded by Huish, the Grand Junction became, in 1846, part of the great London & North Western, a vast railway empire stretching from London to Carlisle.

Another military man involved in early Scottish railway history is Captain J W Coddington, of the Royal Engineers, who was secretary and general manager of the Caledonian from 1847 to 1852. He was one of the few railway managers to have been on both

Glasgow & South Western Railway telegraph instrument, known as a 'block instrument'. This is used for communication between signal boxes, to ensure that a train is not allowed to enter a section of track until the signalman has been advised that the previous train has left the section.

Tablet for single line working. The 'tablet' is the driver's authority to proceed. In the signal boxes at each end of a single line section there are electrically-linked tablet instruments. One tablet can be removed from either instrument but both instruments then lock automatically until the tablet has been replaced in one of the instruments. Although largely superseded by other methods, this system is still used on some lines.

sides of the fence, having been one of the Board of Trade railway inspectors from 1844 to 1847.

A strong general manager like Huish could, if the chairman and directors let him, become more involved in power politics than in running the railway. The transition to manager as operator rather than politician is illustrated by the appointment of John Connacher as general manager of the North British in 1891. Connacher had started his working life in 1861, with the Scottish Central Railway at Perth, when he was 16 years old. Along with the Scottish Central he was absorbed by the Caledonian in 1865 and had to move to Glasgow. Within a few weeks he concluded that there was no future for him there and he moved south to the Cambrian Railways and in time rose to the position of general manager. He returned to Scotland as general manager of the North British, to replace John Walker who had died suddenly after seventeen years in the job. Most of Walker's time was taken up by company politics and in particular, fighting the Caledonian. In seeking a replacement, the directors were quite clear in their minds that they wanted 'a man to work the railway, not one to engineer its policy'. Fighting the Caledonian was not quite over, but it was becoming ever more apparent that it was not good for the finances of either company.

The major railway companies became the largest business enterprises of their time. Their operation required vast amounts of paperwork and to process this an army of clerks was essential. First-hand accounts by railwaymen are not common and those by men from the clerical staff are particularly scarce. The autobiography of Joseph Tatlow is of special interest because it is a personal account of a career which progressed from junior clerk to general manager, in the course of which he served with English, Scottish and Irish companies. He was born in 1851. His father worked for the Midland Railway, whose head office was in Derby and in 1867 Joseph began work with the company as a junior clerk at a salary of £15 per annum. One of the innovations in the office work of the late 1860s was the introduction of shorthand or

'Pitman's phonography'. The Midland encouraged staff to attend classes in this and indeed offered financial inducements to those who became proficient. Tatlow reached the required standard and became a 'shorthand writer', taking down letters at dictation. The typewriter was not in use by then, so that letters had still to be copied out from the shorthand notes.

After five years with the Midland Railway he decided that he was not progressing as he had hoped. It was time for a move, and at the end of 1872 he came to Scotland as private clerk to the stores superintendent of the Caledonian Railway. What his initial salary was he does not say, but after six months service with the Caledonian he was earning £100 a year, and became an income tax payer, at the rate of 3d (1p) in the pound on £40 of his salary. He records that at first he felt quite proud of being a taxpayer!

Late in December 1874 Tatlow heard quite by chance from a friend who worked for the Glasgow & South Western Railway that their general manager was looking for a principal clerk and

Model of a Caledonian Railway express locomotive of 1859, with 8ft (2438mm) diameter driving wheels. Made in the NMS workshop.

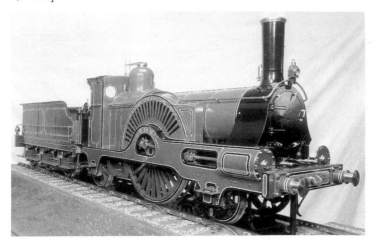

that there was felt to be no suitable internal candidate. He applied and was appointed at a salary of £120. As a springboard to greater things in the railway world there were few better places than the kind of job Tatlow had landed. In the general manager's office he had the opportunity to learn much regarding the legal, financial, commercial and operating aspects of railways and he came to know many of the senior officers from other companies. After ten stimulating years with the Glasgow & South Western, involved in a wide variety of work, he began to feel restless again. Seeing little prospect of promotion he left Scotland for Belfast, to take up the post of general manager of the Belfast & County Down Railway at an annual salary of £500, a very substantial sum at that time. The remainder of his working life was to be spent in Ireland.

Tatlow was a keen observer of life and characters in the railway world, and his remarks on the people he encountered are full of interest. He comments that the clerks he met when he joined the Midland Railway had not been brought up as clerks. They had drifted into the service from other walks of life. These men, and they would all be men, were the first generation of railway clerks, some of whom would have joined the company in its very early years. By the time Tatlow joined the Midland there was already a strong family element in the industry, son following father.

Railways also needed engineers, both civil and mechanical. It is worth noting here that where the term 'engineer' is used without qualification, this almost always refers to the civil engineer. Civil engineering tends to be regarded, at least by civil engineers, as the most senior of the branches of engineering. In railway work, this was so in another sense also, because in any project the civil engineers were involved right at the beginning, while the mechanical engineers and operating staff had to wait until the line was nearly ready for opening. Engineers who were concerned with the construction sometimes stayed on as employees of the railway company. George Graham's long service as engineer of the Caledonian has already been mentioned. Similarly the first engineer of the North British, James Bell, had

Wheel changing at Dunfermline Upper shed, around 1925.
Some improvisation has been required as suitable lifting equip-
ment has not been available at the shed. The breakdown crane
from Thornton Junction had to be called in.

worked with John Miller on the building of the line. He was succeeded by his son who was also James.

The new railway also needed men for track maintenance work and a few of the navvies who had worked on the construction, and felt that the time had come to settle down, might be offered jobs. The maintenance men, known in Scotland as 'surfacemen', worked in small gangs each with responsibility for a section of track. They carried out the routine work needed to keep it in good order. This would involve such jobs as lining up kinks in the rail, packing ballast under any sleepers which had settled, tightening loose bolts on fishplates (which held the lengths of rail together), and keeping drains clear.

The directors of a new company often had only a limited knowledge of the practicalities of running a railway. Before they appointed their own staff they were heavily dependent on their consulting engineers for advice on all technicalities, not just civil engineering matters. The example of the North British Railway, whose engineers were Grainger and Miller, illustrates this well. In addition to organizing the design and construction of the railway itself, John Miller had to prepare a specification for the 26 locomotives required, and advise on a suitable builder. The order was placed with R&W Hawthorn of Newcastle in 1844, nearly two years before the opening of the line. By the time the North British was in a position to appoint a locomotive superintendent, only a few months before the opening, more than half of the locomotives had been delivered and the remainder were not far behind. The man recommended by John Miller for the job, was Hawthorn-trained Robert Thornton, who was induced (presumably by more money) to move from the Edinburgh & Glasgow Railway where he had been in charge of the depot at Haymarket.

In Scotland there was at that time no locomotive building firm comparable to Hawthorn's but the engineering industry was well established in other fields. Many Scots who rose to the top as locomotive engineers later in the nineteenth century received their training and early experience with firms engaged in other branches of engineering. The careers of the Stirling family show this clearly. The Rev Robert Stirling, Minister of Galston, Ayrshire, had a great interest in engineering and in 1816 he patented a novel form of hot-air engine. His brother, James, was also intended for the ministry but became a mechanical engineer instead. He was apprenticed to the Glasgow marine engineer, Claud Girdwood and then became engineer at Deanston mill, in Perthshire, one of the largest cotton-spinning mills in Scotland. From there he moved to Dundee, as manager of the Dundee Foundry Company. The main business was the manufacture of machinery for the textile industry, and marine engines. A few locomotives were also built for local railways in the 1830s.

Dunfermline Upper on a Sunday in July 1926. The locomotives based here mainly worked the heavy coal traffic of the area. They are a mixture of ex-North British Railway types and the new designs of the London & North Eastern Railway.

From 1837 to 1843, Patrick, son of the Rev Robert Stirling, was apprenticed to his uncle James at the Dundee Foundry. For the next ten years he had a very varied career, much of it in marine engineering, but including a brief spell as locomotive superintendent of the small Caledonian & Dunbartonshire Junction Railway. Then in 1853 he settled finally into railway work, becoming the locomotive superintendent of the Glasgow & South Western Railway. He moved to the Great Northern Railway, based at Doncaster, in 1866 and remained in office there until only a few days before his death in 1895 at the age of 75. By the time Patrick's brother, James, started on his career things were changing. He was fifteen years younger and by then locomotive engineers tended to be trained as such more or less from the start. After a few years with the millwright in his home village of Galston, he was apprenticed to his brother, with the Glasgow & South Western Railway. He had a year with the Manchester locomotive-building firm of Sharp Stewart, after which he returned to the Glasgow & South Western. He became works manager at the Kilmarnock workshops and succeeded as locomotive superintendent when Patrick moved to Doncaster. James too eventually succumbed to the attractions of England, becoming locomotive superintendent of the South Eastern Railway, based at Ashford in Kent, in 1878. Both brothers, but Patrick in particular, came to be regarded as among the greatest of the nineteenth-century British locomotive engineers.

The Drummond brothers Dugald and Peter, born 1840 and 1850, came from a railway background, their father being employed by the North British Railway. They had their initial experience in general engineering, both serving their apprenticeship with the Glasgow firm of Forrest & Barr who turned out a wide range of products. Among them were railway cranes, including large breakdown cranes. Part of their business was the hiring of plant to contractors, including locomotives. After his time with Forrest & Barr, Dugald Drummond spent two years with a firm of railway contractors. From there he went to the Cowlairs works of the Edinburgh & Glasgow Railway, subsequently working in various capacities in the locomotive departments of the Highland Railway, and the London, Brighton & South Coast Railway. In 1875 he returned to Scotland as locomotive superintendent of the North British Railway. He moved to the Caledonian in 1882 and finally, in 1895, he went south again as locomotive superintendent of the London & South Western Railway. His work had a profound effect on British locomotive design in general, not just on the companies with which he was directly connected. Some of the locomotives which were running in Scotland in the 1960s clearly showed their Drummond ancestry. When Peter Drummond finished his

Caledonian Railway express nearing the Beattock summit, with the driver out on the front end attending to lubrication.

apprenticeship he immediately went to work with his brother on the London, Brighton & South Coast Railway and then followed him to the North British and Caledonian. In 1896 he became locomotive superintendent of the Highland Railway, moving in 1912 to the Glasgow & South Western.

The post of locomotive superintendent, later known as chief mechanical engineer, was a key one. He was responsible for all aspects of the provision of the locomotive power to run trains, and often also for rolling stock forming the trains. This involved the design of new equipment, manufacture in the company's own works or purchase from outside contractors, repair and maintenance. The actual day-to-day operations came under the locomotive superintendent also. It was a job which needed technical and administrative ability of a high order, together with a certain strength of character. Both the Stirlings and the Drummonds certainly had that.

New railways, having found locomotives and a locomotive superintendent, had to find drivers and firemen. The North British is once again a good example: having recruited its superintendent from the Edinburgh & Glasgow, it obtained the engine crews from the same source, by the simple expedient of offering more money. With a singular lack of tact but, understandably as there was little money in the kitty, as soon as the vacancies had been filled the wages were reduced. This sowed the seeds of a simmering dispute over wages and working hours which finally erupted in 1850. As a result the North British was left with virtually no engine crews. Their places were filled by a mixture of men from the repair shops, who at least knew something about locomotives, and totally inexperienced men recruited from outside. Train speeds were halved while the raw recruits were learning the job. Contemporary press opinion was that there was not much to the job anyway:

> Engine driving is a job which almost any man may learn in a few days. Engine drivers are a parcel of weavers, waiters, barbers or bakers, men who know as much about the construction of the machine they drive as a carter knows about the anatomy of a horse.

As in most labour disputes in the nineteenth century, it was the employer who had the greater power and the crews eventually went back on the company's terms.

Fortunately for the travelling public such haphazard methods of recruitment were soon superseded. Youths were taken on as engine cleaners and by seniority and experience became firemen and then drivers. Training was largely 'on the job' but progress through the various grades was dependent on being able to demonstrate the required standard of knowledge, both of the workings of the locomotive and of the operating procedures and rules. This system was used all over Britain and survived substantially unchanged until the end of steam.

While the senior management and engineering staff were very likely to move from company to company in furtherance of their

Portrait of a North British locomotive, crew and shed staff, at Haymarket, before 1914. SEA

careers, there was little or no opportunity for more lowly grades to follow suit, although they had to move from place to place to fit the company's needs or to have any chance of promotion. The career of one such efficient and committed 'railway servant' (as distinct from 'officer'), John Graham Kirkpatrick, would be similar to that of tens of thousands, up and down the British railway system. He was born near Dalbeattie in 1870, which was on the Glasgow & South Western system. At the age of eleven he went to work full time on a farm. Two years later he got a job with the railway as van boy on town collection and deliveries, until he was regarded as being old enough to take charge of a cart or van. This emphasizes something which is often overlooked, that even at their peak, the railways were heavily dependent on horse-drawn road transport for local cartage. Feeling that this was something of a dead-end, he managed to transfer to the job of passenger porter. From there a line of possible promotion ran through to station inspector, but with each promotion there was a change of station and consequent house move. He became a charge-hand porter at Castle Douglas and then foreman at Dumfries. Finally he was promoted to station inspector at Stranraer Harbour, from where he retired at the age of 65 after 52 years service to the Glasgow & South Western and its successor the London Midland & Scottish Railway. And the reward for all these years as a railway servant, and public servant too, was an *ex-gratia* pension of ten shillings (50p) a week. His grandson wrote of him 'His was not a notable life but it was a noble one.' That is a fitting epitaph for him and many, many like him in railway service.

Work on the nineteenth-century railways had one major advantage over many other industries: a measure of job security, provided of course one obeyed all the rules. This goes some way towards explaining why, despite the fact that the hours were long and wages often small, labour relations in the industry were better than might have been expected. There was also the knowledge that even after the first unions had been established with difficulty, the companies held the power and were prepared to use it to crush any signs of unrest. The year 1890 and the early part of 1891 saw the most widespread troubles in the industry up until then, culminating in several strikes. The most serious, and ultimately the ugliest, took place in Scotland. Most of the goods guards, drivers, firemen and signalmen on the Caledonian, North British and Glasgow & South Western, belonging to the Scottish Amalgamated Society of Railway Servants, came out on strike on 21 December and traffic virtually came to a standstill. The strike began to collapse on the Glasgow & South Western before the end of the year and it was all over by the end of January.

The issue was excessive hours, rather than money. This was a long-standing grievance and one backed up by the Board of Trade railway inspectorate in numerous reports on accidents involving staff who had been on duty for unreasonably long hours. This was by no means a purely Scottish issue. In the report on one particular accident which resulted in the death of a goods guard employed by an English company, the inspector wrote, 'The booked hours were too long, while the actual hours worked were beyond all reason.' The poor man had been on duty for more than 22 hours and the driver of the locomotive which killed him for almost 24 hours! A week in the working life of one North British goods train driver exemplifies the problem. He was paid by the trip, rather than by hours worked. His regular daily job was supposed to take ten hours, and for ten hours only was he paid. During the week beginning 18 January 1890 he worked a total of 90 hours but received payment for only 60 hours. Excessively long hours had become the rule, rather than something which resulted from exceptional circumstances. The increase in traffic following the opening of the new Forth Bridge in March 1890 made matters worse. Even the passenger trains were horrendously late and goods trains would appear to have been shunted into sidings and forgotten about.

The companies were determined to concede nothing to the strikers. One weapon was eviction from company houses and when the Caledonian did just that at Motherwell on 5 January 1891, with military assistance, there was a major riot which resulted in serious damage to the station and the destruction of a signal box. The one gain from the strike was the setting up of a Parliamentary Select Committee which began in March 1891 to hear evidence regarding overwork. The Committee seems to have been convinced by what it heard that there was indeed a problem of excessive hours. But swayed by the views of the company managements that government interference would render it virtually impossible for them to run the railways, the Committee's recommendation that hours should be reduced was vague and woolly. Still it was a first and much needed step. The companies did not make it easy for their staff to give evidence to the Committee. On 7 April 1892, only a few months after John Connacher had started in his new job as general manager of the

North British, he and three Directors of the Cambrian Railways were in the House of Commons being admonished by the Speaker, for contempt of the House, in that they dismissed John Hood, a stationmaster, because he gave evidence to the Select Committee in the previous year. They were fortunate; there were those in the House who thought that a prison sentence would have been more appropriate.

7 Reshaping

At the end of the nineteenth century the railways had a virtual monopoly of inland transport. It was a monopoly subject to government regulation but a monopoly nonetheless. There was still some competition from waterways and coastal shipping for certain types of traffic and in some parts of the country there was an element of competition between individual companies. Early in the twentieth century the situation began to change, slowly at first but with increasing speed until, by the middle of the century, the whole commercial basis of the railway system had been fundamentally altered.

The introduction of the electric tramcar was perhaps the first indication that change was on the way. It was the extensive system operated by Glasgow Corporation which produced the most serious effect on the railway companies' revenues but wherever electric tramways were introduced, the takings from short-distance passenger rail services were adversely affected. Glasgow's first electric cars began operating in 1898 and the entire system was electrically worked by 1902. Within a few years the network extended far beyond the city boundary, from Airdrie in the east to Paisley in the west. Costly new railway lines such as the Glasgow Central Railway and the earlier City & District, through Queen Street, lost a great many short-distance passengers to the new mode of transport. The suburban service run by the Glasgow & South Western Railway between Govan and Springburn, via the City of Glasgow Union Railway, was an early casualty. It was withdrawn completely at the end of September 1902. Almost completed suburban stations on a new Caledonian line between Paisley and Barrhead, in Renfrewshire, were never used. The competing Glasgow & South Western service lasted only from 1902 to 1907.

The first motor vehicles were also just appearing on the scene and their effect on revenue was quickly felt. As early as 1902 the Highland Railway was experiencing a significant reduction in the number of local first-class passenger journeys. However, the Great North of Scotland Railway was quick to see that motor vehicles were potentially useful as well as being a threat and in 1904 it began to run bus services as feeders to the railway.

During World War I the railways came under Government control. Traffic levels were much higher than usual and maintenance was cut to the absolute minimum. As the end of the war approached, thought began to be given to what should happen to the railway system after hostilities ended. The one thing which quickly became clear was that it could not, in its run-down state, revert to the pre-war company organization with any hope of financial viability. The Minster of Transport in the immediate post-war period was Eric Geddes. Born in 1875, he was the son of a Scottish civil engineer who had been much involved in railway building in India. Young Geddes worked for a time in America, and then in India, but returned to Britain in 1904. He joined the staff of the North Eastern Railway where he rose with great speed. Early in the war he was attached to the Ministry of Munitions and this was the start of a rise in Government service even more rapid than that of his career with the North Eastern. In 1917 he became First Lord of the Admiralty. Shortly after the end of the war, Lloyd George offered him the choice of staying on at the Admiralty or taking charge of a projected new ministry which was intended to reorganize all transport services into an integrated system under Government supervision. He chose the latter. However, the Government's enthusiasm for nationalization waned and the Ministry of Transport which was set up in 1919 under Geddes was something less than had originally been planned. Almost its sole task now was the reorganization of the railways and their detachment from wartime controls.

There was, however, one other job the new Ministry had to deal with, the settlement of compensation claims made by the railway companies. This brought him into conflict with William Whitelaw, the strong chairman of the North British Railway. Whitelaw was born at Coatbridge in 1868. His family was connected with the leading iron firm of William Baird and Company, but William was not interested in joining the firm. After a brief political career he decided that railway management was the thing for him. He joined the board of the Highland Railway in 1898 and became chairman in 1902. He became a director of the North British in 1908 and chairman in 1912, after resigning the chairmanship of the Highland. As the war neared its end Whitelaw was one of those firmly convinced that the pre-war situation of the railway companies could not be restored. Of more immediate concern was the question of compensation for the run-down state into which wartime operational needs, under government control, had brought the railway. Not surprisingly the government was as reluctant to pay compensation to the railways as it was to provide the promised 'homes fit for heroes'. As Whitelaw put it, the government had placed the company in a quagmire and was now leaving it to sink or swim. Eventually, however, after a long legal battle and what also became a personal battle between Geddes and Whitelaw, compensation of £9,790,000 was paid to the North British.

Short of state ownership, the widely accepted solution to the problems of the post-war railways was some form of large scale amalgamation. The Ministry of Transport's view was that one of the products of this process should be a Scottish group. Whitelaw

Newcastleton station early in the first World War. The departing train was taking army volunteers. The railway reached Newcastleton in 1862. SEA

More or less from the start railways made provision for the conveyance of private carriages, using waggons known as carriage trucks. The same service was available in later years for motor car owners. This picture is from the late 1920s or early 1930s, but the location is not known. SEA

was sure that the financial position of the Scottish companies would be so weak that this was not a workable option. He mobilized Scottish business opinion and succeeded in persuading the government that the Scottish companies should be linked to the corresponding English networks. The grouping was effected on this basis by the Railways Act of 1921 which came into operation in 1923. The numerous railway companies in England, Scotland and Wales were brought together in four large groups. The London, Midland & Scottish (LMS) incorporated the Caledonian, Glasgow & South Western and Highland. The North British and Great North of Scotland became part of the London & North Eastern (LNER). In addition there were the Great Western and Southern Railways, neither of which had any

Scottish connections. The Chairmanship of the LNER passed to William Whitelaw, a position which he retained until 1938.

The managements of the newly-created companies did not have their troubles to seek. In addition to rebuilding desperately run-down railways they soon had to face a major new competitor, the bus. After 1918 the number of bus operators grew at a great rate and their services began to take a serious amount of traffic from the railways. In a short time a few large companies became dominant. Among them were W J Thomson's Edinburgh-based Scottish Motor Traction Company (SMT) and Walter Alexander and Sons of Falkirk. From 1928 railway companies were allowed to take shares in bus operators and they quickly did so. Both the LMS and LNER acquired a 25% holding in the SMT Company. Then in 1929 the two railways and the SMT acquired financial interests in Walter Alexander & Sons. This was the precursor of a major reorganization of the Scottish bus industry in the wake of the 1930 Road Traffic Act, which for the first time brought in regulation of bus services. The larger operators, including Alexander's, amalgamated with the SMT which became the holding company for the group, with the LMS and LNER as major shareholders. Within the group a series of non-competing operating companies was set up covering most of Scotland between them.

The expanding bus operations, and the development of freight services by road which occurred at the same time, affected rail services. By 1930 the closure of stations and complete lines was under way. Many rural stations built in the nineteenth century were miles from the places whose names they bore and the railways could not compete with buses and lorries offering virtually a door-to-door service. Cost reductions were necessary and redundancies unavoidable. Although times were hard it was not all gloom. There was considerable expenditure on new equipment, although whether it was as much as was really needed is doubtful. Both the LMS and LNER introduced new high-speed Anglo-Scottish services to mark the coronation of King George VI in

1937. At the other end of the scale, the LNER tried hard to improve the economics of lightly used lines by the introduction of steam railcars, which were cheaper to operate than locomotive-hauled trains. Many of the new cars were put to work in Scotland with some success.

In 1939, before they had fully recovered from the effects of the earlier conflict, the railways once more found themselves operating in wartime conditions under government direction, with high levels of traffic and minimum maintenance. With the election of a Labour Government in 1945 the railways were, this time, in line for nationalization. The Transport Act of 1947, which came into effect in the following year, put not only railways under state control but long-distance road haulage, coach and bus undertakings, inland waterways and docks also. Under the Act, the British Transport Commission was brought into being with overall responsibility for the whole range of nationalized transport undertakings. Railways were run by the Railway Executive, and similar executive bodies were set up for the other modes of transport. Operationally the railways were divided into six Regions, including a Scottish Region which took over the lines of the LMS and LNER in Scotland.

In the years immediately after the war traffic was heavy, largely because there were few alternatives, at least for longer journeys. This initial high level of traffic could not be sustained, however, because of the growing competition from private cars, buses, lorries and airlines. In addition to market-place changes the railways had to cope once more with organizational change. The Transport Act of 1953 denationalized road haulage, abolished the Railway Executive and left a reorganized British Transport Commission directly responsible for the railways. Under the new organization a £1,200 million modernization plan was

Cleaning gang, almost all women, at St Margaret's shed, Edinburgh in the late 1940s. The locomotive is a wartime-built 'austerity' design. SEA

announced in 1955. The most visible element of this was the replacement of steam locomotives by diesel and electric power, but there were many other important but less conspicuous parts of the plan, such as signalling and track improvements and drastic reconstruction of marshalling yards and freight terminals. From the point of view of the Scottish passenger one of the early fruits of modernization was the widespread use of diesel multiple-unit trains. The first of these started running on the Edinburgh-Glasgow route in 1957. Another was the electrification of the Glasgow suburban lines from 1960.

Modernization did not produce the hoped-for financial viability and Britain's railways were once again reorganized. The British Transport Commission was replaced in 1963 by the

A minor operational mishap. A considerable number of people, including a small child, are wondering how to get it back on the rails! This is a locomotive of the North British Railway, built about 1908, and it looks fairly new in the photograph. The location is possibly near Newburgh, Fife. SEA

British Railways Board, under the chairmanship of Dr Richard Beeching, formerly technical director of Imperial Chemical Industries. There then appeared a far-reaching report, *The Reshaping of British Railways*, often referred to as the Beeching Plan. For the first time it was fully recognized that a Victorian railway network could not meet twentieth-century transport needs without fundamental reshaping. Many essential things, such as rural and local passenger services, and movement of small consignments of freight, which the railways had provided in the nineteenth century, could now be better done by the motor vehicle. The nub of the report was the need to identify the things which railways could still do as well as or better than other modes of transport, develop the services for these and withdraw from the

A feature of the 1955 plan for the modernization of Britain's railway system was the replacement of steam locomotives by diesel and electric power. But there were problems with the transition from steam to diesel and the locomotives proved troublesome in service.

others. No report of this nature can produce all the right answers but it was abundantly clear that 'no change' was not a viable option. Without some such reappraisal the future would have been bleak indeed for rail transport in Britain. Unfortunately the popular conception of the 'Reshaping' report is that it was wholly a negative document, concerned only with line and station closures. The positive features have largely been ignored.

Scotland's railways have indeed been reshaped. The network is much reduced but where the trains still run the service provided is, more often than not, the fastest and most frequent there has ever been. In recent years some new stations have been built and old ones reopened. New equipment has come into service to replace the trains built under the 1955 modernization plan. The railways are continuing to change as they must, because the Scottish economy is not static and consequently the transport needs are also changing. In the biggest upheaval since nationalization in 1948, British Rail is now being prepared for a return to private ownership. The extent to which this will provide practical solutions to the perceived problems of Britain's railways remains to be seen.

FURTHER READING

BAGWELL, Philip *The Railwaymen: the History of the National Union of Railwaymen*, London 1963

BRUCE, W Scott *The Railways of Fife*, Perth 1980

COLEMAN, Terry *The Railway Navvies*, London, 1965

DOTT, George *Early Scottish Colliery Waggonways*, London 1947

KENNEDY, D *The Birth and Death of a Highland Railway*, London 1971

MCGREGOR, John *100 Years of the West Highland Railway*, Glasgow 1994

MCKENNA, Frank *The Railway Workers 1840-1970*, London 1980

MARSHALL, John *A Biographical Dictionary of Railway Engineers*, Newton Abbot 1978

MARTIN, Don *The Garnkirk & Glasgow Railway*, Kirkintilloch 1981

MARTIN, Don *The Monkland & Kirkintilloch and Associated Railways*, Kirkintilloch 1995

MITCHELL, Joseph *Reminiscences of My Life in the Highlands*, 2 volumes 1st published 1883, reprinted Newton Abbot 1971

NOCK, O S *The Caledonian Railway*, London c1961

PERKINS, John *Steam Trains to Dundee*, Dundee 1975

STEPHENSON LOCOMOTIVE SOCIETY *The Glasgow and South Western Railway*, London 1950

TATLOW, Joseph *Fifty Years of Railway Life in England, Scotland and Ireland*, London 1920

THOMAS, John *A Regional History of the Railways of Great Britain*. Volume 6 *Scotland: the Lowlands and the Borders*, 2nd edition revised by A J S Paterson, Newton Abbot 1984

THOMAS, John *The North British Railway*, 2 volumes, Newton Abbot 1969 & 1975

THOMAS, John *The Tay Bridge Disaster*, Newton Abbot 1972

THOMAS, John and TURNOCK, David *A Regional History of the Railways of Great Britain*. Volume 15 *The North of Scotland*, Newton Abbot 1989

VALLANCE, H A *The Great North of Scotland Railway*, Newton Abbot 1989

VALLANCE, H A *The Highland Railway*, Newton Abbot 1985

WELBOURN, Nigel *Lost Lines: Scotland*, Shepperton 1994

GLASGOW, MAY, 1922.

Golfers' Train
Leaving St Enoch
Station

Golfing Resorts

on the

GLASGOW & SOUTH-WESTERN RAILWAY.

DAVID COOPER.
GENERAL MANAGER.

PLACES TO VISIT

Space allows mention of only a few of the places where material relating to the history of Scotland's railways can be found. In addition to those listed many smaller museums, libraries and record offices have objects, documents and other sources of information such as files of local newspapers. There are also societies devoted to the study of the history of each of the five major pre-grouping Scottish railway companies and they too have collections of archive material.

Much of the early railway infrastructure survives, some disused but a lot of it still in regular use. The trains may be modern but the earthworks, bridges and many of the stations which can be seen on any train journey illustrate the Victorian origins of the system. This 'industrial archaeology' of the railway repays careful study. A few examples, some still in use and some disused, are listed below, but there are many others.

Museums, documentary sources and preserved railways
Aviemore, Inverness-shire: the southern terminus of *the Strathspey Railway*. Steam services run to Boat of Garten on the former Highland line.

Bo'ness, West Lothian: the terminus of the steam-worked Bo'ness & Kinneil Railway, and the headquarters of the *Scottish Railway Preservation Society*. The Society has the most comprehensive collection of historic railway material in Scotland.

Golfers once provided a significant amount of revenue for the railways. Special trains were run, as this Glasgow & South Western Railway booklet of 1922 shows, and stations were built to serve some courses. With increasing car ownership this traffic was soon lost.

Brechin, Angus: a fine example of a Caledonian Railway branch line terminus. Steam trains run to Bridge of Dun.

Dalmellington, Ayrshire: the *Scottish Industrial Railway Centre* at the former Minnivey colliery houses many examples of the locomotives and other equipment used on the railway systems of industrial sites such as collieries and steelworks.

Dundee: at the *McManus Galleries* there is material relating to early railways around Dundee.

Edinburgh: the *Royal Museum of Scotland* has railway items and additional material will be included in the new Museum of Scotland which is under construction.

Edinburgh: most of the surviving records of the Scottish railway companies are at the *Scottish Record Office*, West Register House, where they are available for consultation.

Glasgow: several full-size locomotives from the old Scottish railway companies are displayed at the *Museum of Transport*, along with numerous smaller items and models.

Glasgow: the *Mitchell Library* has extensive photographic collections, including that of the North British Locomotive Company.

Glenfinnan, Inverness-shire: the station building has been restored and now houses a museum dealing with the history of the West Highland Railway and its extension to Mallaig.

Kirkintilloch, Glasgow: the *William Patrick Library* has an important collection of phiotographs and other material, relating particularly to the railways of the Monklands.

Maud, Aberdeenshire: this former Great North of Scotland Railway junction station is now a museum.